NATIONAL NETWORK FOR MANUFACTURING INNOVATION PROGRAM
ANNUAL REPORT

Executive Office of the President
National Science and Technology Council
Advanced Manufacturing National Program Office

February 2016

About this Document

This is the annual report for the National Network for Manufacturing Innovation (NNMI) Program.[1] As required by the Revitalize American Manufacturing and Innovation (RAMI) Act of 2014,[2] the report describes the NNMI Program's major accomplishments in the most recent one-year period. To establish a standardized reporting period, September 30 is used as the cut-off date for this and future annual reports (September 30, 2015 for this report). These annual reports provide a mechanism to assess the program's progress in meeting the goals stated in the *NNMI Program Strategic Plan*. See Appendix A for descriptions of the federal organizations that actively participate in the program and that contributed to this report.

Copyright Information

This document is a work of the U.S. Government and is in the public domain (see 17 U.S.C. § 105).

---

[1] This report refers to the "Network for Manufacturing Innovation Program" as the "National Network for Manufacturing Innovation Program" or "NNMI Program" since the latter is well-established and recognized in practice and because the purpose of the Administration's NNMI initiative launched in 2012 is essentially the same as the Congressionally-authorized program enacted in 2014.

[2] Consolidated and Further Continuing Appropriations Act, 2015, Title VII – Revitalize American Manufacturing and Innovation Act of 2014, codified at 15 U.S.C. § 278s(f)(2)(C).

Interagency Working Team Participants

The individuals listed below dedicated considerable time and expertise to write and produce the *2015 NNMI Program Annual Report*.

Advanced Manufacturing National Program Office
Dannielle Blumenthal
Rolf Butters
Kevin Chou
Lisa Jean Fronczek
Frank W. Gayle
Shreyes N. Melkote
Michael F. Molnar *(Sponsor)*
Frank Pfefferkorn
Margaret Phillips
Robert Rudnitsky
Michael Schen
Gloria Wiens

Department of Commerce
Mary Ann Pacelli
Erin Sparks
Dana Smith
Ben Vickery

Department of Education
Gregory Henschel *(Sponsor)*
Robin Utz

Department of Defense
Michelle Bezdecny
John Christensen
Ralph Day
Scott Frost
Mark Gordon
Abhai Kumar
Scott Leonard
John Olewnik
Adele Ratcliff *(Sponsor)*

Department of Energy
David Hardy
Rob Ivester
Mark Johnson *(Sponsor)*
Steve Nunez
Dev Shenoy
Mark Shuart
Andrew Steigerwald

National Aeronautics and Space Administration
Frank Ledbetter
LaNetra Tate *(Sponsor)*
John Vickers

National Science Foundation
Bruce Kramer *(Sponsor)*
Mihail C. Roco

U.S. Department of Agriculture
Todd Campbell *(Sponsor)*
Hongda Chen
William Goldner
World Nieh
Valerie Reed
Samuel Rikkers
Cynthia West

Message from the Secretary of Commerce

The National Network for Manufacturing Innovation (NNMI) Program is a critical component of this Administration's competitiveness agenda and of our commitment to creating more sustainable economic growth by keeping the United States on the cutting edge of innovation, research, and manufacturing.

Advanced manufacturing offers great potential across the broad landscape of the U.S. economy. Between one-third and one-half of the economic growth in the United States can be attributed to technological and scientific innovation. At the same time, creating new commercial products and processes from research and engineering developments originating in the nation's universities and research laboratories can be difficult, expensive, and time-consuming.

In a global economy driven by lightning fast collaboration, America's innovators cannot afford to operate in isolation from one another. The NNMI public-private partnership breaks down silos between the U.S. private sector and academia to collaborate on taking industry-relevant technologies from lab to market over the near term. This initiative has been developed as an interagency effort, a business effort at all levels of industry, and an academic effort to assert America's leadership role in advanced manufacturing.

The network has nine manufacturing innovation institutes funded by the Department of Defense or the Department of Energy, each in various stages of development. Together, the institutes have grown their membership rolls to more than 800, initiated 147 research and development projects, contributed to workforce development with a variety of educational programs, and stimulated the growth of the regional ecosystem for manufacturing innovation in their particular technology focus areas.

The manufacturing innovation ecosystems established by the institutes in the NNMI Program are creating the sorely missing collaborative infrastructure and critical mass of intellectual might and technology resources needed for our nation to innovate, produce, compete, and build prosperity for the American people.

Even at this early stage of program development, the accomplishments of the established institutes are promising—proving that a team-based approach to innovation yields tangible results. At the Department of Commerce, we are as excited to lead the network to its fullest potential as we are optimistic about America's manufacturing future.

Our commitment to manufacturing innovation sends an unmistakable message to all: America is "open for business."

Penny Pritzker
U.S. Secretary of Commerce

This page intentionally left blank.

Executive Summary

Substantive improvements in the health, robustness and innovative capacity of the U.S. manufacturing sector have an unrivaled ability to boost the nation's global economic competitiveness. Investments in the advanced manufacturing technologies being pursued by the National Network for Manufacturing Innovation (NNMI) Program can have widespread and meaningful impact. They not only revitalize existing industries, but help to create entirely new ones as well.

This report provides an overview of the origin, primary goals, and early progress of the NNMI Program in the first year since the passage of the Revitalize American Manufacturing and Innovation (RAMI) Act of 2014.[2] Working together, the NNMI partners have undertaken a historic effort to support industry in establishing the ecosystems or industrial commons that will better enable innovators to develop the specific manufacturing technologies, processes, and capabilities needed to advance promising early stage technological inventions that continually arise in America's unparalleled basic research sector such that they can be scaled-up and commercialized by U.S. manufacturers.

The work that led to the NNMI Program began in earnest in 2011, when the President's Council of Advisors on Science and Technology (PCAST) recognized the need for a concerted, whole-of-government effort to increase the footprint of applied research and development in advanced manufacturing. Crossing the gap from inventing things to actually making them, at scale, was often beyond the risk limit of the nation's largest manufacturing firms, and still more difficult for the more than 300,000 small and mid-sized enterprises (SMEs) that provide the bulk of U.S. manufacturing output.[3]

Much attention is focused today on "disruptive" innovation, in which new technologies empower start-up companies that create novel consumer products outside the boundaries of traditional manufacturing. Advanced manufacturing is crucial to realizing the full potential of such disruptive innovation. It can improve the efficiency of new manufacturing processes, making U.S. industry a stronger competitor on the global stage. It can provide rapid adaptability in design and production, enabling U.S. industry to respond effectively to changing customer needs. And it can improve sustainability by reducing energy use and decreasing waste.

Whether for disruptive innovation or incremental improvements of existing manufacturing, the value of advanced manufacturing is immense. The interplay of advanced manufacturing with products generates creative synergy: innovative products made possible by new technology often require equally innovative manufacturing methods; conversely, new manufacturing processes may be the key to commercializing a product design previously thought unrealizable.

Engaging researchers, designers, engineers, and manufacturers in a productive public-private partnership is the fundamental motivation behind the NNMI Program, a PCAST recommendation that has now been turned into reality. The NNMI Program includes a number of distributed manufacturing innovation institutes, located at regional hubs strategically poised to bring together manufacturing expertise and

---

[3] *Report to the President on Ensuring American Leadership in Advanced Manufacturing*, Executive Office of the President, President's Council of Advisors on Science and Technology, June 2011,
www.whitehouse.gov/sites/default/files/microsites/ostp/pcast-advanced-manufacturing-june2011.pdf.

relevant research capabilities. The essence of each institute is collaborative and typically pre-competitive work on the development and production scale-up of promising technologies.

Led by a non-profit organization and supported by a consortium of industry, academia and national laboratories, each institute is funded jointly by private and public sources. The institutes engage large manufacturers and SMEs, universities, and local community colleges. A full-time staff of scientists, engineers, and industry experts manage a shared infrastructure for technology development. Institutes feature industry representatives from large and small companies sharing their manufacturing best practices. A range of education and workforce development activities is also integral to the design of these institutes.

In these ways, the institutes serve as a bridge between the industry, research, and education communities, spreading technological knowledge as widely as possible throughout the commercial sector. They bring marketplace experience and pre-competitive technical challenges to the academic community and engage young people in the promise of tomorrow's manufacturing landscape.

By the end of September 2015, seven institutes had been established. They are:

- America Makes, the National Additive Manufacturing Innovation Institute, Youngstown, Ohio, focusing on additive manufacturing/3D Printing technologies.
- Digital Manufacturing and Design Innovation Institute, Chicago, Illinois, focusing on integrated digital design and manufacturing.
- PowerAmerica—The Next Generation Power Electronics Manufacturing Innovation Institute, Raleigh, North Carolina, focusing on wide bandgap semiconductor based power electronics.
- Lightweight Innovations for Tomorrow, Detroit, Michigan, focusing on lightweight metals manufacturing technology.
- Institute for Advanced Composites Manufacturing Innovation, Knoxville, Tennessee, focusing on advanced fiber-reinforced polymer composites.
- AIM Photonics—American Institute for Manufacturing Integrated Photonics, Rochester, New York, focusing on integrated photonic circuit manufacturing.
- NextFlex, America's Flexible Hybrid Electronics Manufacturing Institute, San Jose, California, focusing on the manufacturing and integration of semiconductors and flexible electronics.

In addition, two manufacturing innovation institute competitions are underway—one focused on revolutionary fibers and textiles and one focused on smart manufacturing, advanced sensors, and process controls.

Regionally grounded yet nationally connected and impactful, these institutes leverage the talent of their surrounding ecosystem and support the collective vision of U.S. global leadership in advanced manufacturing.

Table of Contents

Interagency Working Team Participants iv
Message from the Secretary of Commerce v
Executive Summary vii

List of Tables x

List of Figures x

Introduction 1
   A National Need 1
   Manufacturing Innovation Institutes – Initial Efforts 2
   Formal Establishment of the NNMI Program 3
      Vision, Mission, and Goals 4
      Institutes 5
      Network 8

Program Performance 9
   Department of Defense Institutes 9
      America Makes, the National Additive Manufacturing Innovation Institute 10
      Digital Manufacturing and Design Innovation Institute (DMDII) 16
      LIFT: Lightweight Innovations for Tomorrow 19
      AIM Photonics 24
      NextFlex: America's Flexible Hybrid Electronics Manufacturing Institute 25
   Department of Energy Institutes 26
      PowerAmerica—The Next Generation Power Electronics Manufacturing Innovation Institute 27
      Institute for Advanced Composites Manufacturing Innovation (IACMI) 30

NNMI Program Coordination 33
   Network Meetings and Collaboration 33
   Network Functions, Governance, and Interagency Coordination 33
   Supporting Small Businesses Across the Network 34
   Strategic Plan and Annual Report 34
   Public Clearinghouse of Information 35
   Funds Expended 35
   Future NNMI Plans 35

Conclusion 37

Appendix A. Federal Sponsors of the NNMI Program 39

Appendix B. Abbreviations 42

List of Tables

Table 1. Level 1 Network-Level Functions of the NNMI Program 34

List of Figures

Figure 1. Workshop on the Design of the NNMI, January 2013 2
Figure 2. Chronology of Events Associated with NNMI Program Formation 4
Figure 3. NNMI Program Vision, Mission, and Goals 4
Figure 4. NNMI Program Goals 5
Figure 5. Institute Innovation Ecosystem 6
Figure 6. Institute Ecosystem Process Flows Generate Benefits that Incentivize Ongoing Participation...11
Figure 7. DMDII's Assets for Sustainable Value 17
Figure 8. Launch of Virtual Reality Experience during Manufacturing Day 2015 21

Introduction

A National Need

The United States has long thrived as a result of its ability to manufacture goods and sell them in global markets. Manufacturing activity has supported our economic growth, leading the nation's exports and employing millions of Americans. Historically, the manufacturing sector has driven knowledge production and innovation in the United States by employing the scientists, engineers and technicians who drive most research and development.

As highlighted by the 2011 President's Council of Advisors on Science and Technology (PCAST) report,[4] the nation's historic leadership in manufacturing is at risk. Manufacturing as a share of national income has declined significantly since the 1970s, and our leadership in producing and exporting manufactured goods is under threat. The report concluded that the U.S is lagging behind other high-wage nations such as Germany and Japan in research and development linked to the manufacturing sector. The report also underscored the importance of maintaining technological superiority in advanced manufacturing for sustaining U.S. national security and global economic competitiveness.

A follow-on 2012 PCAST report,[5] prepared by the Advanced Manufacturing Partnership (AMP) Steering Committee, noted that although the United States excels at basic science and invention, the commercial and economic rewards that stem from early stage discovery-based research are often realized much later in the innovation life-cycle—specifically, in areas that involve manufacturing scale-up and commercialization. The report highlighted the technical and financial limitations of the U.S. private sector, particularly small and mid-sized manufacturing firms, in translating promising early stage research into cost-effective, high-performing domestic manufacturing capability and new products.

In addition, federal investment in applied research and development relevant to advanced manufacturing is substantially less than the federal investment in early stage basic scientific research. The future ability of the United States to invent and innovate new products and industries, provide high quality jobs to its citizens, and ensure national and economic security depends upon the ability to bridge the scale-up gap and enable rapid transition of advanced technologies to the manufacturing sector.

To address the national needs highlighted above, PCAST, in their 2011 report,[4] recommended launching a concerted, whole-of-government effort in advanced manufacturing, coordinated by the Executive Office of the President (EOP), involving all departments and agencies that help support the nation's advanced manufacturing innovation ecosystem. The requirements identified for such an effort were: 1) coordinated federal support to industry and academia for applied research on new technologies; 2) public-private partnerships to advance such technologies through pre-competitive consortia that tackle major cross-cutting challenges; and 3) shared facilities and infrastructure to help small and mid-sized enterprises (SMEs) improve their manufacturing capabilities and products to compete globally. PCAST also

---

[4] *Report to the President on Ensuring American Leadership in Advanced Manufacturing*, Executive Office of the President, President's Council of Advisors on Science and Technology, June 2011,
www.whitehouse.gov/sites/default/files/microsites/ostp/pcast-advanced-manufacturing-june2011.pdf.

[5] *Report to the President on Capturing Domestic Competitive Advantage in Advanced Manufacturing*, Executive Office of the President, President's Council of Advisors on Science and Technology, July 2012,
www.whitehouse.gov/sites/default/files/microsites/ostp/pcast_amp_steering_committee_report_final_july_17_2012.pdf.

recommended increased support for workforce training in advanced manufacturing technologies to ensure that the nation has the highly skilled workforce needed to attract, retain, and expand advanced manufacturing in the United States. PCAST further recommended[5] the establishment of a national network of public-private partnerships (institutes) designed to foster innovation ecosystems for advanced manufacturing.

These recommendations served as the basis for the Administration's initial efforts to establish manufacturing innovation institutes under the National Network for Manufacturing Innovation (NNMI) initiative.[6]

Manufacturing Innovation Institutes – Initial Efforts[7]

Acting on the recommendations of the PCAST, the President directed the Departments of Defense (DoD), Energy (DOE), and Commerce (DOC), the National Science Foundation (NSF), and the National Aeronautics and Space Administration (NASA) to exercise existing spending authorities to jointly launch a proof-of-concept pilot manufacturing innovation institute focused on a technology topic of importance to the core missions of the respective federal agencies.

In August 2012, America Makes, the National Additive Manufacturing Innovation Institute—focused on additive manufacturing/3D printing—was established in Youngstown, Ohio. It is a public-private partnership that initially consisted of 40 member companies, 9 universities, 5 community colleges, and 11 non-profit organizations.

Figure 1. Workshop on the Design of the NNMI, January 2013

Encouraged by the early success of the pilot institute, and based on extensive public and private sector input gathered through a federally-sponsored Request for Information and a series of five "Designing for Impact" regional workshops, the Administration proposed the creation of the NNMI initiative to revitalize and accelerate U.S. advanced manufacturing by catalyzing the development of new technologies, educational competencies, manufacturing processes, and products via public-private partnerships involving federal, state, and local agencies, the private sector, and academia.

---

[6] *National Network for Manufacturing Innovation: A Preliminary Design*, Executive Office of the President, National Science and Technology Council, Advanced Manufacturing National Program Office, January 2013.
[7] The initial efforts described in this section represent the Administration's activities carried out under the NNMI initiative prior to passage of the RAMI Act of 2014.

INTRODUCTION

The Administration also launched competitions for three additional manufacturing innovation institutes focused on Next Generation Power Electronics Manufacturing (funded and overseen by DOE), Digital Manufacturing and Design Innovation (funded and overseen by DoD), and Lightweight and Modern Metals Manufacturing (funded and overseen by DoD).

In January 2014, the Next Generation Power Electronics Manufacturing Innovation Institute (subsequently known as PowerAmerica)—focused on enabling the next generation of energy-efficient, high-power wide bandgap semiconductor-based electronic chips and devices—was announced at North Carolina State University in Raleigh, North Carolina. It is a public-private partnership that initially consisted of 18 companies, and 7 universities and national laboratories.

In February 2014, the Digital Manufacturing and Design Innovation Institute (DMDII)—focused on accelerating the development and commercialization of digital manufacturing and design technologies—was established in Chicago, Illinois as a public-private partnership that initially consisted of 41 member companies, 23 universities and national laboratories, and 9 other organizations.

Also in February 2014, the Lightweight and Modern Metals Manufacturing Institute (now known as Lightweight Innovations for Tomorrow, or LIFT)—focused on facilitating the development and transition of advanced lightweight and modern metals manufacturing capabilities to the U.S. industrial base—was established in Detroit, Michigan as a public-private partnership that initially consisted of 36 member companies, 12 universities, and 18 other organizations.

Formal Establishment of the NNMI Program

The initial institutes created by the Administration under its 2012 NNMI initiative provided an important basis for Congress to authorize the Secretary of Commerce to establish the "Network for Manufacturing Innovation Program" through passage of the Revitalize American Manufacturing and Innovation (RAMI) Act of 2014[8] on December 16, 2014. This report refers to the congressionally authorized "Network for Manufacturing Innovation Program" as the "National Network for Manufacturing Innovation Program" or "NNMI Program" since the latter is well-established and recognized in practice and because the purpose of the Administration's NNMI initiative launched in 2012 is essentially the same as the congressionally authorized program created in 2014. The Secretary of Commerce has tasked the Advanced Manufacturing National Program Office (AMNPO) at the National Institute of Standards and Technology (NIST) to oversee and carry out the NNMI Program. A chronology of events leading up to the formation of the NNMI Program is shown in figure 2.

Activities and accomplishments of the NNMI Program consist of two essential components: 1) those associated with (or within) a single "Center for Manufacturing Innovation" (referred to in this report as an "institute"),[9] and 2) those associated with and coordinated across the entire "Network for Manufacturing Innovation" (referred to in this report as the "network").

---

[8] Consolidated and Further Continuing Appropriations Act, 2015, Title VII – Revitalize American Manufacturing and Innovation Act of 2014, codified at 15 U.S.C. § 278s.

[9] The terms "Centers for Manufacturing Innovation", "Manufacturing Innovation Institutes", "Institutes for Manufacturing Innovation", and "Clean Energy Manufacturing Innovation Institutes" have all been used in different framing documents to refer to institutes in the NNMI Program.

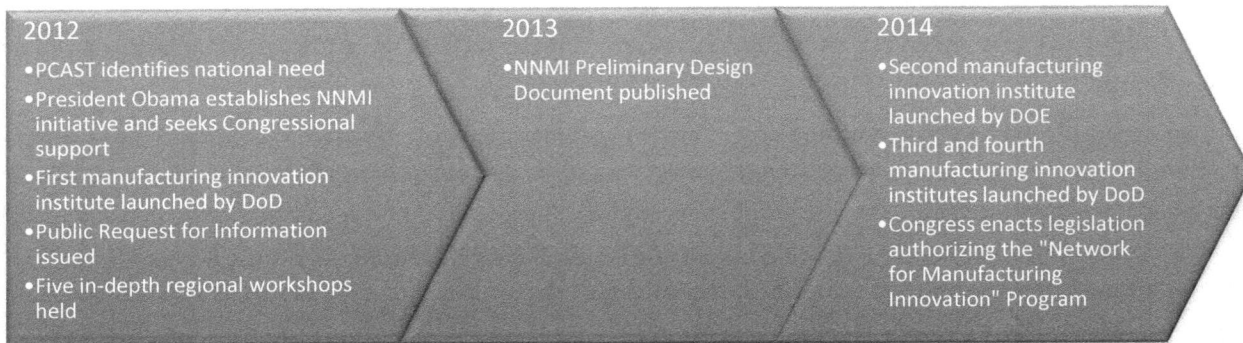

Figure 2. Chronology of Events Associated with NNMI Program Formation

Vision, Mission, and Goals

The NNMI Program seeks to address the complex, manufacturing related technology transition challenges faced in the activity space that falls between early stage basic research and technology deployment in manufacturing. To provide ongoing focus and guidance for its stakeholders toward that end, a vision, a mission, and a set of program goals were developed over the course of 2015, as shown in figure 3.

Figure 3. NNMI Program Vision, Mission, and Goals

INTRODUCTION

The NNMI Program goals, presented in figure 4, are elaborated more fully in the NNMI Program's first strategic plan.

Competitiveness: Increase the competitiveness of U.S. manufacturing		
Technology Advancement: Facilitate the effective transition of innovative technologies into scalable, cost-effective, and high-performing domestic manufacturing capabilities	**Accelerate Manufacturing Workforce Development:** Accelerate the development of an advanced manufacturing workforce	**Sustainability:** Support business models that help institutes become stable and sustainable

Figure 4. NNMI Program Goals

Individual institutes form regional clusters of various architectures and business models as they carry out all of these activities, culminating in the development and maturation of sustainable and nationally impactful manufacturing innovation ecosystems.

Institutes

The institutes represent the essential core of the NNMI Program. As envisioned by the National Science and Technology Council (NSTC),[6] each institute brings together stakeholders such as industry, academia (including universities, community colleges, technical institutes, etc.), federal laboratories, and federal, state, and local governments to form an innovation ecosystem that collaboratively addresses high-risk manufacturing challenges and assists manufacturers in retaining and expanding industrial production in the United States. Members of the innovation ecosystem co-invest with the Federal Government, forming a strong public-private partnership designed to accelerate non-federal investment in the development and deployment of advanced manufacturing capacity in the United States, thereby enhancing the competitiveness of U.S. manufacturing. Figure 5 depicts the main elements of the manufacturing innovation ecosystem created by an institute.

Figure 5. Institute Innovation Ecosystem

A key focus of the NNMI Program is to link innovation and manufacturing. To that end, institute activities include:

- Pre-competitive applied research, development, and manufacturing scale-up projects to reduce the cost, time, and risk of commercializing new manufacturing technologies and improvements to existing technologies, processes, and products.

- Development and implementation of education and training programs.

- Development of innovative methodologies and practices for supply chain integration and introduction of new technologies into supply chains.

- Outreach and engagement with SMEs and large-sized manufacturing firms.

INTRODUCTION

The institutes help transition ideas from the laboratory prototype phase through the manufacturing scale-up phase, i.e., from manufacturing readiness levels (MRLs) 4 through 7.[10] There is a loose relationship between MRLs and technology readiness levels (TRLs). For consistency, the MRL scale will be used throughout the document.

The institutes function with the requirement that non-federal resources must equal or exceed the federal contribution during a five to seven year initial institute establishment period. Institutes are expected to become self-sustaining following the initial federal establishment period.

Each of the seven current institutes was established to address a focused manufacturing technology area with initial funding from lead funding agencies under individual agency statutory authorities and appropriations.

1. America Makes, The National Additive Manufacturing Innovation Institute (additive manufacturing/3D printing), August 2012.[11]
2. DMDII: Digital Manufacturing and Design Innovation Institute (digital manufacturing and design), February 2014.[12]
3. LIFT: Lightweight Innovations for Tomorrow (lightweight metals manufacturing), February 2014.[13]
4. PowerAmerica—The Next Generation Power Electronics Manufacturing Innovation Institute,(wide bandgap power electronics manufacturing), December 2014.[14]
5. IACMI: Institute for Advanced Composites Manufacturing Innovation (fiber-reinforced polymer composites), June 2015.[15]
6. AIM Photonics—American Institute for Manufacturing Integrated Photonics (integrated photonics manufacturing), July 2015.[16]
7. NextFlex, America's Flexible Hybrid Electronics Manufacturing Institute (thin flexible electronic devices and sensors), August 2015.[17]

Competitions for two additional institutes, funded under individual agency statutory authorities, are currently underway.

- Revolutionary Fibers and Textiles Institute for Manufacturing Innovation (award anticipated in Fiscal Year 2016).[18]

---

[10] dodmrl.com/MRL_Deffinitions_2010.pdf.
[11] americamakes.us.
[12] dmdii.uilabs.org.
[13] lift.technology.
[14] www.poweramericainstitute.com.
[15] iacmi.org.
[16] www.aimphotonics.com.
[17] nextflex.us.
[18] grants.gov/web/grants/view-opportunity.html?oppId=276514.

- Clean Energy Manufacturing Innovation Institute on Smart Manufacturing: Advanced Sensors, Controls, Platforms and Modeling for Manufacturing (in solicitation).[19]

These and all future institutes leverage public and private resources to create the physical, digital, and relational infrastructure required for applied research and development in a focused manufacturing technology area.

Network

As a part of the strategy to revitalize American manufacturing, the RAMI Act authorizes the Department of Commerce to establish and convene a nationwide network comprised of the individual manufacturing innovation institutes, which can enhance their impacts and further strengthen America's global competitiveness. For example, as part of the network's activity, institutes would be able to share best practices, amplify a shared vision of manufacturing excellence for the United States, identify and address gaps in America's manufacturing technology base, identify common interests and activities that can help to train the next-generation of skilled workers, transition newly developed manufacturing technologies and processes to the U.S. industrial base, and leverage expertise across multiple disciplines.

The network is expected to provide a variety of functions based on the evolving needs of its members. After much coordination this past year, a draft set of four, network-level (i.e., "Level 1") functions of the NNMI Program were identified by the participating agencies. They are:

- Establish the network.
- Facilitate value-added, intra-network collaboration.
- Foster robust communication between the network (of institutes) and external stakeholders.
- Sustain, strengthen, and grow the network.

A detailed discussion of these network-level functions is contained in the section entitled, "Network Functions, Governance and Interagency Coordination," which appears later in the report.

---

[19] www.energy.gov/articles/energy-department-announces-70-million-innovation-institute-smart-manufacturing.

Program Performance

The major activities and accomplishments of the manufacturing innovation institutes through the end of Fiscal Year 2015 are described in this section of the report.[20] The institutes are organized by their sponsoring federal agency. With the exception of the very recently established institutes, the achievements of each institute are highlighted in the four strategic areas, which map to the goals of the *NNMI Program Strategic Plan*. They include 1) Technology Advancement (Applied Research and Development), 2) Workforce Development, 3) Sustainability, and 4) Innovation Ecosystem Development, to which all of the previous areas contribute.

Each institute works toward strengthening the industrial ecosystem surrounding its advanced technology focus area. In support of the NNMI Program vision, the collective work of the institutes in the four strategic areas serves to achieve the program goals.

Department of Defense Institutes

The Department of Defense (DoD) manufacturing innovation institutes represent a key new investment strategy for the Department and its DoD Manufacturing Technology (ManTech) program. These institutes are designed to overcome the critical challenge faced when new and emerging technologies or technology families that hold strategic promise for the DoD are at risk of stalling or collapse because the existing fragmented and frail innovation ecosystems—particularly those technologies and supporting industrial infrastructure aligned too heavily or exclusively with DoD requirements—are insufficient to sustain the technology's development into scaled production. These new institutes are designed to overcome this industrial challenge through the establishment of robust, public-private partnerships incentivized to create sustainable manufacturing innovation ecosystems focused on technology spaces with both strong defense and promising commercial applications. The DoD has established five such institutes and has one more planned for Fiscal Year 2016.

---

[20] The activities described in this first *NNMI Program Annual Report* reflect efforts since the announcement of each of the existing institutes including those prior to Fiscal Year 2015. Future reports will restrict themselves to a single fiscal year's actions.

America Makes, the National Additive Manufacturing Innovation Institute
americamakes.us

Mission: To accelerate the adoption of additive manufacturing technologies to increase domestic manufacturing competitiveness.

Locations: Main hub, Youngstown, OH; Satellite: El Paso, TX

Established: Awarded, August 2012; Facility opened, October 2012

Consortium Organizer: National Center for Defense Manufacturing and Machining (NCDMM)

Funding: Federal, $55M, including support from DOE ($10M), DOC-NIST ($5M), NSF, and NASA; Non-federal, $55M (planned funding over five years)

Members:[21] 149

> "America Makes is making us all proud. …This is exactly the way it's supposed to be. This is a critical program for U.S. global competitiveness."
>
> —U.S. Secretary of Commerce Penny Pritzker, August 27, 2015

Established in August 2012, America Makes is the first of the institutes. After three years, America Makes has not only brought together the elements of an ecosystem, it has established the process flows that will make it a *sustainable* ecosystem. These process flows, shown in figure 6, generate enough benefit to the participating organizations (industry, academia, government, and others) to incentivize their ongoing participation and investment in America Makes, and to attract new organizations that bring additional benefits and investment to the ecosystem.

---

[21] As of September 30, 2015.

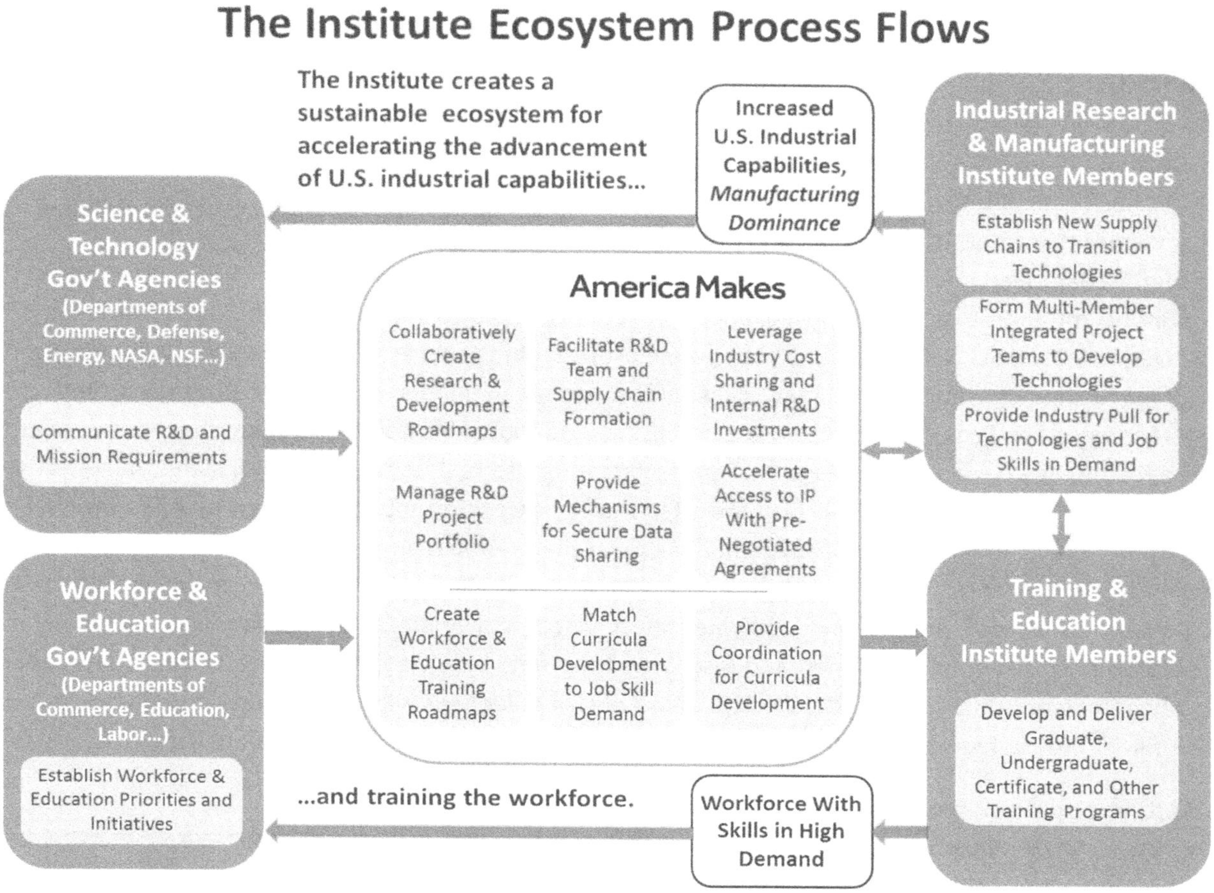

The Institute Ecosystem Process Flows

Figure 6. Institute Ecosystem Process Flows Generate Benefits that Incentivize Ongoing Participation

The process flow begins with a participatory approach to planning. America Makes members and government agency partners involved in science and technology development—such as the Departments of Commerce, Defense, and Energy, NASA, and NSF—identify and prioritize additive manufacturing research and development (R&D) requirements through collaborative roadmapping workshops.

America Makes then works with its members to form teams, leverage investments, manage the research portfolio, share data, and accelerate access to intellectual property. Members form integrated project teams to conduct research and create supply chains to transition technology into production. The result is increased U.S. manufacturing capabilities, moving industry towards achieving *manufacturing dominance*, which is a prerequisite for technical, economic, and defense dominance.

Likewise, America Makes members and government agencies involved in workforce education—such as the Departments of Commerce, Education, and Labor—establish their priorities for workforce development. America Makes works with its members to create roadmaps for workforce development programs, identify job skills that are in demand, and develop curricula matched to those job skills. Members deliver the courses, resulting in a workforce trained for high-demand jobs in the additive manufacturing industry.

Technology Advancement

America Makes has led the creation of an industry-driven strategic roadmap for advancing additive manufacturing technology. The strategy highlights critical technology capability gaps and opportunities for U.S. technological leadership. These advancements include the creation of design tools, novel materials, next generation equipment capable of printing multiple materials, and advanced process controls to reduce manufacturing variability. To achieve these technical advancements, integrated project teams are frequently organized as supply chains, including technology innovators, material and component suppliers, equipment producers, and large system integrators. Project teams form strong business relationships between supply chain partners early in the technology development stage, and projects are tied to real-world applications and market-driven opportunities. This team structure facilitates technology transition into production by ensuring that all requirements are understood at all levels in the supply chain. As just one example, a team led by Northrop Grumman, in partnership with small business and part manufacturer Oxford Performance Materials, demonstrated a high performance polymer as a viable material choice for air and space vehicle applications. This project developed and distributed the first widely-available materials design database for any polymeric additive material, sharing critical design guidelines with the industry and demonstrating the ability to reuse feedstock materials, thereby greatly reducing the overall manufacturing cost.

> **R&D Projects Launched**
> - Project calls have led to 31 industry-driven projects launched thus far.
> - Today, the total R&D portfolio includes 58 projects from all sources.

Workforce Development

America Makes has partnered with several organizations to develop and deliver successful workforce training programs.

- **Deloitte Consulting:** Partnered to create a free 3-hour online course in additive manufacturing business fundamentals. Over 10,000 participants have taken the course so far.
- **Milwaukee School of Engineering & Society of Manufacturing Engineers:** Co-launched the first-ever 3D printing and additive manufacturing certificate program. The program has awarded over 150 certificates as of mid-2015.
- **Private Industry:** Partnered with members to raise private donations that provided desktop 3D printers to over 1,000 schools (K-12) for their Science, Technology, Engineering, and Mathematics (STEM) education programs.

> **Case Studies: Partnering for Impact**
> - 10,000 participants have taken an online course in additive manufacturing business fundamentals.
> - First-ever 3D printing and additive manufacturing certificate training program created and over 150 certificates awarded as of mid-2015.
> - America Makes partnered with members to raise private donations that provided desktop 3D printers to over 1,000 K-12 schools for their STEM education programs.

America Makes is continuing a robust roadmapping process to identify and prioritize future additive manufacturing workforce education and training programs.

Sustainability

Membership Growth
The benefits of joining the America Makes ecosystem is evidenced by growth in membership, breadth of membership, and the level of membership support:

- The institute has reached nearly 150 members with roughly 40% net growth annually.

- Members represent all levels of the additive manufacturing supply chain: large and small businesses, universities, community colleges, economic development organizations, federal laboratories, and government partners.

- Member organizations contribute through annual cash and/or in-kind contributions of $200,000 for Platinum-level membership, $50,000 for Gold-level, and $15,000 for Silver-level.

Small Business
Small businesses are clearly demonstrating the benefits and value they are gaining from being members of America Makes. Part of the value is that America Makes is providing a conduit between larger industry opportunities and small business innovation.

> **About Small Business Membership**
>
> Approximately 50 small businesses have joined the institute thus far, and the institute has an annual retention rate of 88% for small businesses.

- Working with other America Makes members allowed a small business, rp+m of Avon Lake, Ohio, to obtain the necessary AS9100C certification to serve in high-tech supply chains, enabling the company to do business with major companies, including Lockheed Martin, Northrop Grumman, GE Aviation, and Boeing.

- Small business Optomec led a team that developed a hybrid manufacturing system (additive and subtractive manufacturing capabilities) by embedding a modular design, which retrofits any CNC machine tool to be upgraded with 3D printing capability. This development enables machine shops to incorporate additive manufacturing capabilities at 60% less cost than purchasing a new system. This innovation is now commercially available and was launched with significant initial sales, indicating strong customer interest.

- Steelville Manufacturing is entering the market for 3D printed tools, making aerospace components as a supplier to Boeing. Steelville Manufacturing's market entry was enabled by one of the first America Makes technology development projects—Sparse-Build Rapid Tooling by Fused Deposition Modeling for Composite Manufacturing and Hydroforming—led by the Missouri University of Science and Technology.

Public-Private Partnership
Recognizing that America Makes is a national asset for additive manufacturing technology development, federal agencies have begun partnering with America Makes to address specific research and development interests that align with both the industry-driven strategic roadmap and agency needs. These projects add to the shared portfolio of research within the institute and provide further opportunities for the public and private sectors to pool resources to address important technical goals.

Member Feedback

All 24 members participating in a roundtable discussion with Secretary of Commerce Penny Pritzker at America Makes on August 27, 2015 strongly praised America Makes. A recurring message from the members was that America Makes is fostering a trusted member network, or ecosystem. Some comments:

- Lockheed Martin: "America Makes is the strongest public-private partnership I've seen in my 32-year career."

- General Electric: A representative said that the institute enables collaboration that would otherwise not be possible.

- Youngstown State University: "America Makes is having an extraordinary impact on our region. Youngstown has a new confidence. Local government is working with universities and local businesses more. Youngstown State University is proud to have America Makes nearby, and it is now part of our student recruiting story."

- Lorain County Community College: America Makes is "invaluable for helping community colleges to understand what workforce skills are in demand."

Innovation Ecosystem Development

Members are demonstrating how important the ecosystem is by investing directly in its creation and operation.

- Raytheon is sponsoring an employee with additive manufacturing expertise to devote half of her time as an adjunct staff member of America Makes, and she is serving in industry leadership roles as Chair of both the Roadmapping Advisory Group and the America Makes Executive Committee.

- American Society of Mechanical Engineers is sponsoring a full-time one-year Fellow who is assigned to the Workforce Education and Outreach efforts at America Makes.

- Deloitte Consulting is sponsoring employees who are helping America Makes to develop technology commercialization processes and to evaluate America Makes' economic impact.

Private Investment

America Makes is drawing significant private investment related to additive manufacturing into the region.

- Alcoa is building a $60 million R&D Center in New Kensington, PA. It will include a state-of-the-art additive manufacturing center focused on feedstock materials, processes, product design and qualification.[22] The investment will advance the development of proprietary metal powders engineered specifically for additive manufacturing.

How the Institute is Extending its Reach

In August 2015, America Makes launched a pilot satellite center at the University of Texas at El Paso. It will implement the full slate of America Makes institute capabilities—technology development, technology transition, and workforce education and training.

---

[22] www.onlineamd.com/alcoa-expand-additive-research-pittsburgh-090915.aspx#.VfHeOBHBzRb.

- **General Electric** is investing $32 million to build a 125,000-square foot Advanced Manufacturing Facility near Pittsburgh, PA. The center will help their business units develop and implement 3D printing, as well as other innovative technologies.[23]

On a national level, several large business members report they are re-aligning their internal R&D investments with the America Makes technology investment roadmap.

---

[23] www.post-gazette.com/business/development/2014/11/13/General-Electric-planning-advanced-manufacturing-here/stories/201411130182.

Digital Manufacturing and Design Innovation Institute (DMDII)
dmdii.uilabs.org

Mission: To digitize American manufacturing.

Location: Chicago, IL

Established: Awarded, February 2014; Facility opened, May 2015

Consortium Organizer: UI LABS

Funding: Federal, $70M; Non-federal, $106M (planned funding over five years)

Members: [24] 140+

Technology Advancement

> "The widespread availability of design and modeling tools will encourage a new wave of innovation and creativity as the number of people who participate in the marketplace increases…the unprecedented access to powerful software tools, models, and the means of production will serve to democratize the entire manufacturing process."
>
> —Dr. Joe Salvo, Manager, Complex Systems Laboratory, GE Global Research.

After just over a year, the institute has organized a far-reaching consortium and provided the foundations for collaboration.

To this end, DMDII has convened an internal team of subject matter experts and organized multiple committees to shape the strategic direction of the institute. DMDII's Technical Advisory Committee, which consists of member representatives from industry, academia, and government, has created a digital technology roadmap and yearly strategic investment plans that translate the technology roadmap into research projects.

DMDII has also defined an R&D project management process that has enabled collaboration amongst the consortium members and resulted in the selection of 32 projects.

Workforce Development

DMDII launched and convened the DMDII Workforce Development Advisory Committee to build a multi-year strategic roadmap for the institute's workforce development pillar. The committee identified three major components needed in the area of digital manufacturing and design:

- Define the domain and workforce skills needed to succeed.
- Educate through the development of content, promotion, and train-the-trainer initiatives.
- Develop and publish initial use cases and thought pieces on digital manufacturing.

These components led to the development of initiatives that are laying the foundation for project calls to create courses for an online specialization in digital manufacturing and design as well as creating a digital manufacturing skills taxonomy with job profiles.

---

[24] As of September 30, 2015.

Sustainability

> Collaboration among its 140 members is the essence of the DMDII and the institute has defined an R&D project management process that strongly enables it.
>
> *"At the project proposal meeting I met Dan Mulligan of the University of Delaware. We spent the whole night talking about ways to develop a project, what it might be structured like, and how it would look. We really just stayed up all night and wrote it all out until 6 a.m. and then flew back home...and then started pulling the team together."*
>
> —Neil Gupta, Green Dynamics co-founder and Chief Technology Officer

Membership Growth

After just over one year, DMDII has signed over 140 members, 100 of whom are from industry.

Business Model

The institute has laid the foundations for a sustainable business model that will enhance the competitiveness of American manufacturing. Highlights include:

- All members have aligned with one membership agreement that complies fully with DMDII's own cooperative agreement with the DoD.
- Members are pursuing a single, comprehensive technology roadmap that sets forth the infrastructure for collaborative research projects.
- The institute and its members are following a structured R&D project management process for collaborative selection of projects.

Figure 7 illustrates the assets DMDII has built to competitively differentiate itself as a premiere innovation model for industry, academia, government, and other non-profit entities.

Figure 7. DMDII's Assets for Sustainable Value

Novel Strategy

One of the pillars of DMDII's sustainability strategy is the Digital Manufacturing Commons (DMC), a software platform that can be used to share and access manufacturing data and analysis tools. DMDII has been able to gather experts in digital manufacturing to participate in a DMC Advisory Committee. The institute announced its first project call for the DMC in July 2015 and conducted a project workshop at the DMDII facility that attracted virtual and on-site participation from potential contributors to the platform.

Innovation Ecosystem Development

DMDII has deployed a multi-pronged strategy to develop the regional ecosystem, improve local competitiveness, and reach SMEs. At an affordable $500, DMDII's Tier 3 industry membership has enabled over 100 small and mid-sized businesses to join the institute as members.

The City of Chicago and the State of Illinois combined to commit $16.5 million to convert a long-time vacant manufacturing facility into a state-of-the-art hub for digital manufacturing. Additionally, member companies have loaned more than $3 million in equipment and multiple software systems for the manufacturing lab.

DMDII serves as the center of a broader ecosystem, convening institute members to enable collaboration and learning. The establishment of the institute has led to multiple new nearby real estate developments, which are turning an old manufacturing district into a vibrant innovation community. The local investment has also catalyzed a series of related activities. These activities include local workforce development initiatives and the creation of a working group to assess the feasibility of developing the space adjacent to the DMDII facility into a new home for technology incubators that could support spin-offs of the institute.

Furthermore, DMDII's facility has acted as a hub for engagement, hosting over 2,000 visitors within its first six months of existence, including prospective members, veterans, foreign delegations, the U.S. Secretary of Commerce, representatives of multiple student organizations, local Chicago manufacturing initiatives, technology startups, and business incubators and accelerators.

Outside of the facility, DMDII has engaged over 380 organizations through roadshows held at Rolls-Royce, at the State University of New York at Buffalo, and in Rochester, the Quad Cities, Wisconsin, and Colorado, among others. Finally, as part of a partnership between the University of Illinois and UI LABS called the Illinois Manufacturing Lab (IML) program, DMDII has executed eight projects with small manufacturers in the Illinois region helping them deploy digital technologies and identify significant savings. During the second phase of IML, DMDII will scale solutions for small manufacturers using its open source platform, the DMC. As an outgrowth of the strategy for the IML, DMDII plans to expand its reach by launching satellite chapters. These chapters represent a model that can be expanded nationwide to better reach SMEs.

LIFT: Lightweight Innovations for Tomorrow
lift.technology

Mission: Speed development of new lightweight metal manufacturing processes for products using lightweight metal, including aluminum, magnesium, titanium, and advanced high strength steel alloys. Train workers to use these new processes in factories and maintenance facilities.

Locations: Main hub: Detroit, MI; Satellites: Columbus, OH; Ann Arbor, MI; Worchester, MA; Golden, CO

Established: Awarded, February 2014; Facility opened, January 2015

Consortium Organizer: American Lightweight Materials Manufacturing Innovation Institute

Funding: Federal, $70M; Non-federal, $78M (planned funding over five years)

Members: [25] 82

Technology Advancement

LIFT has kicked-off its technology development activities in the areas of melt and powder processing, thermo-mechanical processing, joining and assembly, coatings, and agile tooling. The institute has held two user-led (both commercial and government stakeholders) ideation sessions, one request for proposals, and two organization-specific projects. These activities evaluated over 200 potential project topics.

> **Number and Types of R&D Activity**
>
> LIFT has approved approximately 31 projects, either initiated or under development, in the areas of melt and powder processing, thermo-mechanical processing, joining and assembly, coatings, and agile tooling. These projects are advancing technologies for warfighting, aerospace, automotive, and other applications.

Technology projects are always:

- Led by an industry partner.
- Composed of participants from the entire supply chain.
- Supported by a technology transition plan as a key element of the full work-plan.
- Augmented by a secondary application in a different industry sector for which the technology could be applied.
- Supplemented by an education and workforce development plan.

---

[25] As of September 30, 2015.

The LIFT team is also actively managing the development and implementation of industry-ready Integrated Computational Materials Engineering (ICME) tools, including standardized practices for tool development as a core competency. These ICME tools will enable industry to reduce the time, cost, and risk of manufacturing technology development and to accelerate the deployment of lightweight metals and processes.

LIFT members have access to the full capabilities of the institute's organizations and equipment to conduct fee-for-service projects. This is a key enabler for taking the technology from MRL 7 through MRL 10.[9]

Workforce Development

> *"Just think—50 energized teachers across middle Tennessee translates to as many as 5,000 middle school and high school students who are making career, education, and training decisions! Imagine we do this again, and increase coverage of the state educators!"*
>
> —Feedback received on LIFT/ASM Materials Bootcamp held in Tennessee

LIFT has built the infrastructure to design and implement workforce education solutions in its five-state region—Ohio, Michigan, Indiana, Tennessee, and Kentucky. This infrastructure is a key to visibility, support, and sustainability for LIFT.

Early Successes

As a learning hub, LIFT's investments in education and workforce development have spanned the continuum of challenges related to the current, critical skills gap. These investments have also created opportunities for generating greater interest in careers in advanced manufacturing, including the skills sets necessary to commercialize and produce using new and emerging technologies, materials, and processes.

Each LIFT education and workforce development initiative meets the tests of sustainability, replicability, and moving to scale, helping to achieve the goal of workforce readiness for advanced manufacturing.

Broad Workforce Development Input

LIFT has created five state LIFT teams spanning Ohio, Michigan, Indiana, Kentucky, and Tennessee. The teams involve 98 education, workforce development, economic development, business, industry, and government officials, to align with LIFT's and the NNMI Program goals, and to design and implement workforce development solutions in their respective states.

Examples of investments include:

- LIFT invested in a "mission-focused" lightweight aircraft design curriculum that is being used in school systems reaching 25,000 students in 22 states; sponsored a Purdue-designed Indy 500 Grand Prix for high school students to engineer, build, test, and market vehicles using lightweight metals; and integrated content on lightweight metals and new technologies in 45 teacher boot camps run by ASM International reaching a thousand teachers this year.

- To rebuild the U.S. advanced manufacturing technical workforce, LIFT partnered with Indiana's Vincennes University in an accelerated machining training program for veterans; invested in the development of industry standards for industrial technology maintenance jobs—one of the most in-demand manufacturing jobs nationally;[26] and launched an apprenticeship learning pathway model for adults in Kentucky's community and technical college system.

- As LIFT rolls out its new technology projects, the institute's education and workforce development infrastructure is preparing new internship models, challenge prizes for university and community college students, and workshops for incumbent workers to integrate new competencies aligned with emerging technologies and processes.

Case Study: A Virtual Reality Experience

To begin creating LIFT as a "learning lab" as well as a technology lab, LIFT developed with Tennessee Tech University a virtual reality experience to immerse students and workers in an automobile assembly line with the challenge to select lightweight metals for parts to build a vehicle with high speed and energy efficiency.

Figure 8. Launch of Virtual Reality Experience during Manufacturing Day 2015

LIFT has developed and implemented advanced manufacturing workforce demand-supply gap analyses, producing bi-monthly reports for each of the five states in its region and for the metalworking industry nationwide.

Sustainability

Membership Growth

LIFT's initial regional focus is in a five-state region along the I-75 corridor: Michigan, Ohio, Indiana, Kentucky and Tennessee. While LIFT's initial focus is in this region, its membership network is growing. Highlights include:

- LIFT has received 82 signed membership agreements covering approximately 25 states.

- Membership includes organizations at all levels of the lightweight metals manufacturing supply chain.

- Technology project teams include participation from the complete manufacturing ecosystem, including large companies, research partners, SMEs, and workforce and education intermediaries.

---

[26] Bureau of Labor Statistics, U.S. Department of Labor, Occupational Outlook Handbook, 2014-15 Edition, Industrial Machinery Mechanics and Maintenance Workers and Millwrights, on the Internet at www.bls.gov/ooh/installation-maintenance-and-repair/industrial-machinery-mechanics-and-maintenance-workers-and-millwrights.htm (visited August 31, 2015).

- To build the talent pipeline and promote employment in advanced manufacturing, the institute is working with 98 education and workforce network organizations representing the five-state region. More than a dozen national programs are now fully engaged in LIFT.

The institute is pursuing multiple paths toward growth by becoming a national resource for sustained development and commercialization of new materials and manufacturing technologies. Highlights include:

- Working to increase its share and quantity of government programs.
- Growing industrial memberships to go beyond its cooperative agreement.
- Developing two fee-for-service projects.
- Licensing intellectual property.
- Initiating a project for the U.S. Army Tank Automotive Research, Development and Engineering Center (TARDEC).

This work complements the pre-competitive projects and allows the companies to advance the technology to their particular application.

Innovation Ecosystem Development

LIFT signed a lease for the institute headquarters in Detroit on July 23, 2014. The ribbon-cutting was on January 15, 2015.

Infrastructure

LIFT Headquarters is located in historic Corktown, one of Detroit's oldest neighborhoods, and has emerged as a vital component of the city's revitalization efforts. The facility was the former home of Mexican Industries, which made plastic moldings for the automobile industry until 2001, when the company filed for bankruptcy and the building was abandoned. To make this facility home, LIFT, the city of Detroit, and the State of Michigan are investing more than $10 million to create a world-class, state-of-the-art innovation and collaboration space for lightweighting technology development and workforce education. In regards to LIFT, Detroit Mayor Mike Duggan states, "What you see here is not just about advancing technology, it's about advancing people. The education and training collaborations will help prepare Detroiters for employment opportunities to design, build, and repair the next generation of lightweight vehicles." In addition, Mayor Duggan extended the city's Innovation District to include LIFT. Detroit's Innovation District is a 4.3 square mile urban research and technology-focused innovation corridor. Paula Sorrell, then Vice President of the Michigan Economic Development Corporation said, "Michigan already has the world's highest concentration of automotive research and development facilities, so it is an ideal location for this collaborative enterprise. We know we will have to work together as government, research institutions, and private companies to grow our manufacturing base, and LIFT will be an important part of that effort."

> **Partnering with NIST MEP**
>
> LIFT has established a new partnership with the local NIST Manufacturing Extension Partnership (MEP) Center, the Michigan Manufacturing Technology Center, as the conduit to SMEs. On July 21, 2015, the first SME members' meeting was held at the institute's headquarters in Detroit, with nearly 40 people in attendance.

To grow the ecosystem and to ensure the participation of the entire supply chain in each project, LIFT is reaching out to SMEs through its member trade associations and plans to have SMEs take part in all projects in the technology portfolio.

A technology help line is being established and facilitated by LIFT's partner, the Michigan Manufacturing Technology Center, a NIST Manufacturing Extension Partnership (MEP) Center. The help line will provide assistance with technical inquiries from institute members and will link members with technical capabilities across the ecosystem.

Chapters

Developing advanced manufacturing methods for metallic components typically requires access to high capital investment processing equipment and advanced materials characterization facilities. In addition to the pilot scale equipment that is being procured for the headquarters high-bay, LIFT has established "core partner sites" where LIFT members will have access to appropriate equipment. These sites include—Southeast Michigan (University of Michigan and Comau LLC); Columbus, Ohio (Ohio State University and EWI®); Worcester, Massachusetts (Worcester Polytechnic Institute); and Golden Colorado (Colorado School of Mines).

LIFT is currently looking to expand its ecosystem into regions south and west of its initial five-state focus.

AIM Photonics: American Institute for
Manufacturing Integrated Photonics
www.aimphotonics.com

Mission: Seek to advance integrated photonic circuit manufacturing technology development while simultaneously providing access to state-of-the-art fabrication, packaging, and testing capabilities for small-to-medium enterprises, academia, and the government; create an adaptive integrated photonic circuit workforce capable of meeting industry needs and thus further increasing domestic competitiveness; and meet participating commercial, defense, and civilian agency needs in this burgeoning technology area.

Locations: Main hubs, Albany and Rochester, NY

Established: Awarded, July 2015

Consortium Organizer: Research Foundation for the State University of New York

Funding: Federal, $110M; Non-federal, $502M (planned funding over five years)

Proposal Members:[27] 124+

Institute Overview
Note: This institute was launched in the second half of calendar year 2015. Therefore, a basic overview is provided.

AIM Photonics will seek to automate the assembly of integrated photonics systems to minimize the touch-labor component, whose high cost has prompted industry to seek offshore production solutions in recent decades.

Other needed integrated photonics manufacturing advancements include a quick-turnaround capability for rapidly developing and prototyping designs.

Headquartered in New York State, with founding academic partners in California, Massachusetts, and Arizona, AIM Photonics will bring government, industry, and academia together to organize the current fragmented domestic capabilities in integrated photonics and better position the United States to compete globally.

Applications
- Ultra-high-speed transmission of signals for the internet and telecommunications.
- New high-performance information-processing systems and computing.
- Compact sensor applications enabling dramatic medical advances in diagnostics and treatment.
- Multi-sensor applications including urban navigation, free space optical communications, and quantum information sciences.

---

[27] At launch. Source: Department of Defense.

NextFlex: America's Flexible Hybrid Electronics Manufacturing Institute
www.nextflex.us

Mission: Pioneer a new era of advanced Flexible Hybrid Electronics (FHE) manufacturing in the United States by: 1) catalyzing a U.S. FHE ecosystem; 2) providing new manufacturing capability to the Department of Defense and industry partners; 3) demonstrating FHE manufacturing through relevant technology demonstration platforms; and 4) educating and training professionals and technicians.

Location: San Jose, CA

Established: Awarded, August 2015

Consortium Organizer: FlexTech Alliance

Funding: Federal, $75M; Non-federal, $96M+ (planned funding over five years)

Proposal Members:[28] 160+

Institute Overview

Note: This institute was launched in the second half of calendar year 2015. Therefore, a basic overview is provided.

Flexible hybrid electronics manufacturing is an innovative process at the intersection of the electronics industry and the high-precision printing industry, with the power to create sensors that are lighter in weight, or conform to the curves of a human body, while preserving the full operational integrity of traditional electronic architectures.

Applications
- Medical health monitoring and personal fitness.
- Soft robotics to care for the elderly or assist a wounded Soldier.
- Lightweight sensors embedded into the trellises and fibers of roads and bridges.

Integrating ultra-thin silicon components—through high-precision handling, printing with conductive and active inks, and printing to integrate on stretchable substrates—flexible hybrid technologies can improve the connectivity of devices through the internet of things.

The FlexTech Alliance, a public-private manufacturing consortium based in San Jose, California, will lead the newest institute to secure U.S. leadership in next-generation bendable and wearable electronic devices. The institute hub location will exploit nodes around the country and will leverage centers of excellence in materials, printing, assembly, and related manufacturing technologies.

---

[28] At launch. Source: Department of Defense.

Department of Energy Institutes

The DOE uses manufacturing innovation institutes to develop energy efficiency and clean energy manufacturing technologies to support the Clean Energy Manufacturing Initiative—a DOE initiative to strengthen U.S. clean energy manufacturing competitiveness and to increase U.S. manufacturing competitiveness across the board by boosting energy productivity and leveraging low-cost domestic energy resources and feedstocks.[29]

The DOE partners with private and public stakeholders to support the applied research, development and deployment of innovative technologies with high potential to improve U.S. competitiveness, save energy, and ensure global leadership in advanced manufacturing and clean energy technologies. To achieve these goals, the DOE partners with industries, SMEs, universities, not-for-profit organizations, and national laboratories to support the translation of scientific breakthroughs to advanced manufacturing.

The DOE has responsibility for managing the system of 17 national laboratories to meet the applied science, technology, and national security mission of the department. As a result and where appropriate, the DOE provides a conduit for coordination between existing capabilities and infrastructure of the national laboratories system and the institutes. In addition, through partnerships with the institutes, national laboratories have the ability to identify new early stage basic science and technology challenges that are most relevant to U.S. manufacturers.

Within the DOE, the institutes are managed by the Advanced Manufacturing Office (AMO) of Energy Efficiency and Renewable Energy (EERE) as Clean Energy manufacturing innovation institutes. Topics for the DOE-led institutes resulted from a series of open and public engagements with industry, including requests for information,[30] public technical workshops,[31] as well as the DOE department-wide Quadrennial Technology Review (QTR),[32] conducted in 2015. Through this process, the DOE has ensured these institutes are aligned with both the manufacturing competitiveness mission of the NNMI Program, and the energy efficiency and clean energy technology mission of the DOE.

---

[29] energy.gov/eere/cemi/clean-energy-manufacturing-initiative.
[30] www.energy.gov/eere/amo/articles/new-request-information-rfi-clean-energy-manufacturing-topic-areas.
[31] energy.gov/eere/amo/downloads/manufacturing-innovation-multi-topic-workshop.
[32] *Chapter 6: Innovating Clean Energy Technologies in Advanced Manufacturing*, Department of Energy, Quadrennial Technology Review: An Assessment of Energy Technologies and Research Opportunities, September 2015, energy.gov/quadrennial-technology-review-2015.

PowerAmerica—The Next Generation Power Electronics
Manufacturing Innovation Institute
www.poweramericainstitute.com

Mission: Develop advanced manufacturing processes that enable large-scale production of wide bandgap semiconductors. These allow electronic components to be smaller, faster, and more energy efficient than semiconductors made from silicon. Help reshape the American energy economy by increasing efficiency in everything that uses a semiconductor, from industrial motors and household appliances to military satellites.

Locations: Raleigh, NC

Established: Awarded, December 2014; Facility opened, January 2015

Consortium Organizer: North Carolina State University

Funding: Federal, $70M; Non-federal, $70M (both planned funding over five years)

Members:[33] 25

Technology Advancement

The Next Generation Power Electronics Manufacturing Innovation Institute has initiated 22 projects since February 1, 2015. Highlights include:

- The largest PowerAmerica project is establishing the infrastructure to validate a wide bandgap semiconductor foundry model for power electronics. This foundry model allows companies of all sizes and universities to repurpose existing silicon foundries to fabricate wide bandgap semiconductor devices. Equipment is being qualified and multiple companies now have projects to fabricate and qualify wide bandgap devices in the PowerAmerica partner foundry.

- An innovative open foundry model builds American capacity to produce wide bandgap power electronic devices at high volume and at lower cost.

- Another project focuses on the qualification and release of high voltage devices and modules to shorten the lead-time of building blocks for smaller, more compact, more robust, energy efficient power electronic systems.

Further, recent interactions with industry and supply chain analysts identified thermal potting compounds used for heat management and electrical insulation as an area of investigation. Private sector discussions have since ensued.

Workforce Development

This institute has created a concentration in Wide Bandgap Semiconductor Electronics for graduate students. This program is designed for industry engineers to adapt their design practices for wide bandgap power electronics as well as provide a pathway for undergraduates to focus their graduate work in an emerging technical area of strategic importance.

---

[33] At launch. Unaudited.

PowerAmerica is also working with NIST's MEP Program and NSF's Advanced Technology Education Program, and is engaged in a series of train-the-trainer programs. Over 100 high school teachers applied for the first Introduction to PowerAmerica session. The initial meetings with high school and community college instructors are being used to shape STEM and community college activities so that they align with existing standards and curricula.

Additionally, the teaching surface mount technology manufacturing line is designed to engage SMEs and is expected to be used in conjunction with the MEP program to transfer best design practices and processes to printed circuit board houses. Local industry should have access to the line for development purposes. PowerAmerica is also discussing with IPC® (association connecting electronics industries) on the development of reliability standards for high-temperature wide bandgap semiconductor-enabled printed circuit boards.

Sustainability

Membership Growth
PowerAmerica has engaged in a variety of economic development activities by working with trade associations and economic development groups in the region, such as the Research Triangle Clean Tech Cluster, the North Carolina Sustainable Energy Association, and the North Carolina Clean Energy Center.

Fee-for-Service
A better understanding of the supply chain in the past six months has led PowerAmerica to develop a "Device Bank" concept. The Device Bank allows wide bandgap devices and power modules produced under government funding to be made available to industry and academia for research and development and for test and evaluation. Further, the device bank concept is structured to increase the availability of state-of-the-art wide bandgap devices that otherwise would be prohibitively expensive for small business or university research.

The open foundry concept is structured to encourage small high-tech business and venture capital to invest in wide bandgap semiconductor device fabrication by lowering the initial capital equipment cost and leveraging the open foundry's production expertise.

Private Investment
As one example, PowerAmerica raised approximately $3.5 million in external financing and identified executive leadership for a power electronics group that wanted to spinout from a larger company. A key underlying reason for this action was to ensure that people and technologies remained in the United States. Although this did not result in the formation of a separate company, it helped validate the desire for a different company to absorb the key individuals from the power electronics group and their intellectual property. PowerAmerica is anticipated to stimulate further investment interest in wide bandgap power electronics. Understanding these market dynamics will be key to maximizing the success of the nation's investment in the NNMI Program.

Innovation Ecosystem Development
In addition to working with trade associations and economic development groups in the region, PowerAmerica is also examining economic development opportunities through its participation in the North Carolina Veterans Foundation (as a board member) and the North Carolina Military Business Center.

Other industry engagements are occurring through Defense Network (DEFNET)—which is a monthly meeting of businesses focused on serving the defense industry—the Small Business Development Center, as well as the North Carolina State University's extension and economic outreach programs.

Institute for Advanced Composites Manufacturing Innovation (IACMI)
iacmi.org

Mission: To lower the overall manufacturing costs of advanced composites by 50 percent, reduce the energy used to make composites by 75 percent, and increase the recyclability of composites to over 95 percent within the next decade. Enable their use for a broader range of products including lightweight vehicles with record-breaking fuel economy; lighter and longer wind turbine blades; high pressure tanks for natural gas fueled cars; and lighter, more efficient industrial equipment.

Locations: Knoxville, TN

Established: Awarded, June 2015; Facility opened, June 2015

Consortium Organizer: University of Tennessee

Funding: Federal, $70M; Non-federal, $180M (both planned funding over five years)

Members:[34] 122+

Technology Advancement

IACMI is expected to serve as an ongoing resource for accelerating industry-led innovation by linking complementary capabilities, common objectives, and industry-honed expertise with an application-focused intellectual property management plan.

Early Successes

- In Colorado, the National Renewable Energy Laboratory (NREL) National Wind Technology Center—an IACMI core facility partner—is working with Colorado State University, the Colorado School of Mines, Iowa State University, and the University of Colorado, along with all of the major wind industry original equipment manufacturers to enhance the U.S. market for wind energy generation technologies.

- In Indiana, Purdue University is a founding partner of IACMI and is leading the research and the development of a computational modelling effort. Immediately subsequent to the establishment of IACMI, Purdue broke ground on a $50 million, 62,000 square foot building focused on advanced composites research and development.

- In September 2015, the institute announced that it is co-locating one of its shared research, development, and demonstration resources at LIFT's Corktown facility in Detroit to drive production-scale composite materials development focused on vehicle lightweighting.

- In Tennessee, IACMI is leveraging investments in the Oak Ridge National Laboratory (ORNL) Manufacturing Demonstration Facility and the Carbon Fiber Technology Facility (CFTF).

---

[34] At launch, the institute had 123 entities committed to membership. The institute is in the process of formalizing the membership agreements.

PROGRAM PERFORMANCE

Technical Goals

The institute's technical goals are framed in terms of specific quantitative objectives to advance the state-of-the-art in fiber reinforced polymer composites manufacturing and beneficial impacts from implementing these advances.

The DOE has charged IACMI to achieve the five-year technical objectives of:

- 25% lower carbon fiber reinforced polymer (CFRP) cost;
- 50% reduction in CFRP embodied energy; and
- 80% composite recyclability into useful products, with these objectives based on a manufactured part.

IACMI is expected to attain the technical objectives through market-focused R&D projects, which should draw upon the resources from one or more of IACMI's application or enabling technology areas, and completed projects that should yield beneficial impacts as private industry invests in commercializing the results of the projects to feed market growth.

Workforce Development

IACMI has assessed workforce development needs in the area of composites as a means to develop effective programs. These programs are expected to seek partners with existing infrastructure for creation and delivery, and to link them to IACMI's R&D activities. Highlights include:

- IACMI has contracted with the Workforce Intelligence Network to conduct an analysis of key occupations in the composites industry for the IACMI states.
- An IACMI partnership with the American Composites Manufacturers Association (ACMA) and the Roane State Community College in Tennessee is developing an online composites technician certificate course for vacuum infusion processing. This pilot is envisioned as the first step in providing a new composites production training curricula that can be made broadly available.
- Building on the success of an onsite internship program for college students at the CFTF, IACMI is partnering with Oak Ridge Association of Universities to establish 15 summer internships for college students, three in each of IACMI's technical focus areas.
- IACMI is further engaging with LIFT and their partner the American Society for Engineering Education to bring together existing educational and training content on lightweighting into a single portal to significantly increase access to relevant training materials throughout the nation. IACMI representatives from ORNL, Dow Chemical, the University of Tennessee, the University of Kentucky, Michigan State University, and Purdue University are participating in this project.

Sustainability

IACMI's membership base includes a collection of SMEs and large multinational companies. Collectively, they represent the entire supply chain. An open solicitation for project proposals encourages participation by SMEs through reduced cost share requirements and explicit consideration for impact on SMEs in evaluation of proposals.

Other member companies and organizations represent complementary areas of nondestructive evaluation, inline monitoring for process control, modeling and simulation for process and component design and optimization, and product testing and certification.

Collectively, these member organizations cover the material life cycle through small, medium, and large volume applications, with an emphasis on high volume manufacturing.

This process, together with member meetings, roadmapping, and strong member communications provides a platform that shares opportunities and encourages participation in collaborative projects.

Innovation Ecosystem Development

> *"MVP is just one company out of many that now have a forum to collaborate on ideas that could revolutionize the way that people use composites in their everyday life and not only save energy and pollution but in a very real way, change the world and how we know it."*
>
> —Testimonial from Magnum Venus Products

IACMI is significantly increasing composite research, development, and deployment capacity within the automotive corridor that spans from Michigan into the Southeast.

- The Indiana Economic Development Corporation is investing $35 million in research equipment and materials to support IACMI and a new composites research facility built on land provided by the City of West Lafayette Redevelopment Commission and supported by $11 million in commitment from the Purdue Research Foundation.

- In Ohio, the University of Dayton Research Institute and the National Composites Center serve as a regional hub for composite materials research and application, including the core application area of compressed gas storage.

- The Kentucky Cabinet for Economic Development and the Kentucky Automotive Industry Association hosted the first Auto Vision Conference to support the automotive manufacturing industry, and IACMI senior personnel served on a panel session for developing lighter, more fuel-efficient vehicles.

NNMI Program Coordination

Network Meetings and Collaboration

The RAMI Act requires that the Secretary of Commerce establish a nationwide network comprised of the individual manufacturing innovation institutes. In the past year, the interagency AMNPO has worked closely with its agency partners to lay the groundwork for establishing and convening this network.

Thus far, the AMNPO has convened four meetings of the institutes, two of which were held prior to the passage of the RAMI Act:

- June 6, 2014, in Detroit, MI.
- December 17, 2014, in Gaithersburg, MD.
- March 18, 2015, in Washington, DC.
- November 4, 2015, in Chicago, IL.

These collaborative, network-wide institute meetings have proven to be productive information sharing and idea generation forums that have included formative dialogue on candidate network-level functions, including which functions could be most useful to institutes, both during the initial phase of standing up an institute and for ongoing operations. This early dialogue helped shape subsequent, more focused work to develop and refine a consensus set of draft network functions (see the section below entitled "Network Functions, Governance, and Interagency Coordination").

Collaboration has also taken place among the federal agency members of the AMNPO, ranging from biweekly meetings for planning the management and coordination of the NNMI Program, to higher level policy decisions for defining and moving the network functions forward.

Network Functions, Governance, and Interagency Coordination

Through the NSTC Subcommittee on Advanced Manufacturing (SAM), high level discussions to coordinate programs in advanced manufacturing across the Federal Government are ongoing, including NNMI Program coordination. The SAM discussions include cross-cutting advanced manufacturing topics that are priority technology areas for multiple federal agencies and issues of network governance.

A multi-agency Network Governance Team was tasked in May 2015 to develop and facilitate foundational aspects of the network. The team developed an action plan consisting of 11 steps and has been working through the identified actions. Of note is a network-level functions plan describing high level (Level 1) and more detailed (Level 2) functions, developed and vetted with both interagency and external stakeholders of the NNMI Program. This includes an extensive Responsible, Accountable, Consulted, Informed analysis to determine how the network will operate. Table 1 lists the four Level 1 network-level functions with amplifying notes.

Table 1. Level 1 Network-Level Functions of the NNMI Program

Function	Amplifying Notes
Establish the network	<ul><li>Focus is on initial network form-up tasks.</li><li>Supports RAMI direction to "convene" the network.</li><li>The need for this function will diminish over time.</li></ul>
Facilitate value-added, intra-network collaboration	<ul><li>Focus is on the network's internal information clearinghouse tasks.</li><li>Includes inter-institute and interagency collaborations and information exchange.</li></ul>
Foster robust communication between the network and external stakeholders	<ul><li>Focus is on the network's external information clearinghouse tasks.</li><li>Includes information exchange and messaging with stakeholder entities that are generally viewed as external to the network.</li></ul>
Sustain, strengthen, and grow the network	<ul><li>Focus is on longer-term network sustainment and growth tasks.</li></ul>

As follow-up to this effort, the SAM NNMI Working Group Leaders are coordinating with the NIST AMNPO, lead funding agencies, and the institutes to develop the Network Charter and "Game Plan" for pursuing the highest priority network functional activities—from among the Level 2 network functions—to undertake in 2016 and beyond.

Supporting Small Businesses Across the Network

Over the course of Fiscal Year 2015, NIST MEP executed an MOU with the DoD. This MOU defines how institutes and MEP Centers can work together to facilitate SME awareness of the NNMI Program and to encourage SMEs to participate in institute R&D planning, participate in institute R&D, and implement and deploy institute R&D results.

As a result, beginning in 2015, a pilot is now underway between NIST MEP, nine MEP Centers (including California, Colorado, Georgia, Idaho, Illinois, Indiana, Kentucky, Ohio, and Texas), and the DMDII. This pilot will develop tools to assess a company's "digital readiness" and provide training to help MEP manufacturing specialists and MEP clients understand the growth and competitiveness implications of digital readiness for manufacturing companies. The pilot will also help DMDII conduct outreach and serve as a means for MEP Centers to educate U.S. manufacturers about digital manufacturing.

Strategic Plan and Annual Report

During this last year, an interagency group led by the AMNPO developed the first strategic plan for the NNMI Program. It is complemented by this *NNMI Program Annual Report*, which each year will describe: 1) the specific activities of the agencies participating in the NNMI Program, and 2) the activities of the network during the previous year in achieving the goals described in the strategic plan.

The *NNMI Program Strategic Plan* was drafted in collaboration with representatives from all of the agencies participating in the NNMI Program, including the DOC, the Department of Education (DOEd), the DoD, the DOE, NASA, NSF, and the U.S. Department of Agriculture (USDA). Initial writing was divided into four writing teams, with two chaired by the DOD, one by the DOE, and one by DOC/NIST. The document was then reviewed for additional input by representatives of the seven funded institutes. It was then reviewed by the Office of Science and Technology Policy (OSTP), the National Economic Council (NEC), and the NSTC.

Public Clearinghouse of Information

The AMNPO continued to maintain and upgrade manufacturing.gov, a national advanced manufacturing portal highlighting the NNMI Program. The portal contains information on every institute, whether established or in the planning phase, as well as other items of interest related to advanced manufacturing and domestic manufacturing competitiveness.

The office also established Twitter and LinkedIn accounts to communicate status updates with the public and to facilitate the reception of feedback from them.

Funds Expended

No funds have been provided by the RAMI Act or other legislation for the DOC to establish institutes at the time of this report. Nominal funds have been expended to comply with legislative reporting requirements, including the preparation of this report.

Future NNMI Plans

Future plans of the NNMI Program include:

1. Continue to build and refine the network, its functions, and procedures to operationalize the functions.

2. Strengthen interagency cooperation and coordination with respect to the NNMI Program.

3. Assist lead funding agencies in their efforts to establish new institutes.

4. Work with interagency team members to further define performance metrics for the network.

5. Continue to build the NNMI brand.

6. Continue to serve as a public clearinghouse for information related to the NNMI Program.

7. Coordinate with the institutes through the lead funding agency personnel assigned to the interagency team to formalize data collection for purposes of identifying best practices, informing the congressionally mandated annual report, and evaluating the performance of the network.

This page intentionally left blank.

Conclusion

The NNMI Program is a historic, transformational public-private initiative aimed at creating robust and sustainable ecosystems enabling the innovative development of manufacturing technologies, processes and capabilities that help close the gap between early stage basic R&D and the scale-up and deployment of technologies in manufacturing. This highly collaborative effort brings together our nation's federal agencies, academic institutions, and businesses of every kind and size with a stake in advanced manufacturing. It is an affirmation of the reality that we live in a rapidly changing world where value is created by teams of skilled professionals working side-by-side, generating practical solutions to pressing problems.

Through the NNMI Program, individual institutes leverage collaboration opportunities to support the goal of developing and transitioning advanced manufacturing technologies to the U.S. industrial base. These efforts are expected to have widespread economic impact through their ability to engender new industries and revitalize existing ones, and to sustain our critical technological advantage in the national security sector. Working together, we can develop new manufacturing processes that enable the production of American inventions and improve the efficiency of current manufacturing processes, thereby making U.S. industry a stronger competitor on the global stage. We can provide rapid adaptability in design and production; enabling U.S. industry to respond effectively to changing customer needs. And we can improve the condition of our planet by reducing energy use and decreasing waste.

Together, the institutes in the NNMI Program are enabling the United States to build a powerful national network that can effectively drive innovation, best practices, and long-term economic growth. In just a few short years, the institutes established by the Departments of Defense and Energy have made meaningful strides toward realizing the purposes of the NNMI Program.

Our institutes send an unmistakable message to all: America is "open for business."

This page intentionally left blank.

Appendix A. Federal Sponsors of the NNMI Program

National Economic Council

The National Economic Council (NEC) was established in 1993 to advise the President on U.S. and global economic policy. It resides within the Office of Policy Development and is part of the Executive Office of the President. The NEC has four principal functions: to coordinate policy-making for domestic and international economic issues, to coordinate economic policy advice for the President, to ensure that policy decisions and programs are consistent with the President's economic goals, and to monitor implementation of the President's economic policy agenda. More information is available at www.whitehouse.gov/administration/eop/nec.

Office of Science and Technology Policy

The Office of Science and Technology Policy (OSTP) was established by the National Science and Technology Policy, Organization, and Priorities Act of 1976. OSTP's responsibilities include advising the President in policy formulation and budget development on questions in which science and technology are important elements; articulating the President's science and technology policy and programs; and fostering strong partnerships among federal, state, and local governments, and the scientific communities in industry and academia. The Director of OSTP also serves as Assistant to the President for Science and Technology and manages the National Science and Technology Council (NSTC). More information is available at www.ostp.gov.

National Science and Technology Council

The National Science and Technology Council (NSTC) is the principal means by which the Executive Branch coordinates science and technology policy across the federal research and development enterprise. A primary objective of the NSTC is establishing clear national goals for federal science and technology investments. The NSTC prepares research and development strategies that are coordinated across federal agencies to form investment packages aimed at accomplishing multiple national goals. The work of the NSTC is organized under committees that oversee subcommittees and working groups focused on different aspects of science and technology. More information is available at:
www.whitehouse.gov/administration/eop/ostp/nstc.

Subcommittee on Advanced Manufacturing

The Subcommittee on Advanced Manufacturing serves as a forum within the NSTC for information-sharing, coordination, and consensus-building among participating agencies regarding federal policy, programs, and budget guidance for advanced manufacturing. Originally chartered in 2012, the Subcommittee seeks to identify: gaps in federal advanced manufacturing research and development portfolio and policies, programs and policies that support technology commercialization, methods for improving the business climate, and opportunities for public-private collaboration. Regarding advanced manufacturing programs conducted by the Federal Government, the Subcommittee engages in the identification and integration of multi-agency technical requirements, joint program planning and coordination, and development of joint strategies or multi-agency joint solicitations.

Advanced Manufacturing National Program Office

Hosted by the Department of Commerce at the National Institute of Standards and Technology (NIST), the Advanced Manufacturing National Program Office (AMNPO) is an interagency team with participation from federal agencies involved in advanced manufacturing. Principal participant agencies currently include the Departments of Commerce, Defense, Education, and Energy, the National Aeronautics and Space Administration, and the National Science Foundation. Established in 2012, the AMNPO reports to the Executive Office of the President and operates under the NSTC on cross-agency initiatives. The office reports to the Secretary of Commerce in its role as the "the National Office of the Network for Manufacturing Innovation Program," also referred to as the "National Program Office," as described by the Revitalize American Manufacturing and Innovation Act of 2014. More information is available at www.manufacturing.gov.

Department of Commerce

The Department's mission is to create the conditions for economic growth and opportunity. As part of the Administration's economic team, the Secretary of Commerce serves as the voice of U.S. business in the President's Cabinet. The Department works with businesses, universities, communities, and the nation's workers to promote job creation, economic growth, sustainable development, and improved standards of living for Americans. Through its 12 bureaus and nearly 47,000 employees located in all 50 states and territories and more than 86 countries worldwide, the Department administers critical programs that touch the lives of every American. The Department's workforce is as diverse as its mission. It is made up of economists, Nobel Prize-winning scientists, Foreign Service officers, patent attorneys, law enforcement officers, and specialists in everything from international trade to aerospace engineering.

Within the Department of Commerce, the National Institute of Standards and Technology is a non-regulatory federal agency founded in 1901. NIST's mission is to promote U.S. innovation and industrial competitiveness by advancing measurement science, standards, and technology in ways that enhance economic security and improve our quality of life. The agency operates primarily in two locations: Gaithersburg, Maryland (headquarters) and Boulder, Colorado. NIST employs about 3,000 scientists, engineers, technicians, and support and administrative personnel, and hosts about 2,700 associates from academia, industry, and other government agencies, who collaborate with NIST staff and access user facilities.

Department of Defense

The Department of Defense (DoD) considers the Defense Industrial Base to be a part of its force structure. It is as essential to national security as its people in uniform and DoD civilians. For nearly 60 years, the DoD Manufacturing Technology (ManTech) Program, overseen by the Office of the Deputy Assistant Secretary of Defense for Manufacturing and Industrial Base Policy, has been the Defense Department's investment mechanism for staying at the forefront of defense-essential manufacturing capability and is a key component of maintaining a healthy and resilient Defense Industrial Base. ManTech focuses on enabling the affordable and timely development, production, and sustainment of defense systems, thereby enhancing the nation's technological edge in a dynamic, diverse, and evolving threat environment. The growing family of DoD-led manufacturing innovation institutes has become a key component of the Department's strategy to fully develop and enable the full-scale production of critical technologies across the Defense Industrial Base.

Department of Energy

Within the Department of Energy, the Advanced Manufacturing Office (AMO) partners with industry, small business, universities, and other stakeholders to identify and support research and development of emerging energy technologies with the potential to enhance energy productivity and efficiency in manufacturing. The AMO focuses on applied research, development, demonstration and deployment of new technologies within the mission space of energy efficiency in manufacturing and platform technologies for the manufacturing of clean energy products. AMO plays a leadership role in the National Network for Manufacturing Innovation by establishing and supporting consortia targeted at achieving this mission. The AMO's investments have high impact, target nationally important innovation at critical decision points for manufacturing, and contribute to quantifiable energy savings. By reducing the life-cycle energy consumption of manufactured goods by 50 percent over 10 years, the AMO works to support the creation of high-quality domestic manufacturing jobs and enhance the competitiveness of the United States.

Appendix B. Abbreviations

ACMA	American Composites Manufacturing Association
AIM	American Institute for Manufacturing
AMNPO	Advanced Manufacturing National Program Office
AMP	Advanced Manufacturing Partnership
CFRP	Carbon Fiber Reinforced Plastic
CFTF	Carbon Fiber Technology Facility (ORNL)
DEFNET	Defense Network (DoD)
DMC	Digital Manufacturing Commons
DMDII	Digital Manufacturing and Design Innovation Institute
DOC	Department of Commerce
DoD	Department of Defense
DOE	Department of Energy
DOEd	Department of Education
EOP	Executive Office of the President
FHE	Flexible Hybrid Electronics
IACMI	Institute for Advanced Composites Manufacturing Innovation
ICME	Integrated Computational Materials Engineering
IML	Illinois Manufacturing Lab
ITAR	International Traffic in Arms Regulations
LIFT	Lightweight Innovations for Tomorrow
ManTech	Manufacturing Technology (DoD)
MEP	Manufacturing Extension Partnership (DOC/NIST)
MOU	Memorandum of Understanding
MRL	Manufacturing Readiness Level
NASA	National Aeronautics and Space Administration
NEC	National Economic Council (EOP)
NIST	National Institute of Standards and Technology (DOC)
NNMI	National Network for Manufacturing Innovation
NREL	National Renewable Energy Laboratory
NSF	National Science Foundation
NSTC	National Science and Technology Council
ORNL	Oak Ridge National Laboratories
OSTP	Office of Science and Technology Policy (EOP)
PCAST	President's Council of Advisors on Science and Technology
RAMI	Revitalize American Manufacturing and Innovation
R&D	Research and Development
SAM	Subcommittee on Advanced Manufacturing (NSTC)
SME	Small and Mid-sized Enterprise
STEM	Science, Technology, Engineering, and Mathematics
TARDEC	Tank Automotive Research, Development and Engineering Center (DoD)
TRL	Technology Readiness Level
USDA	U.S. Department of Agriculture

Advanced Manufacturing National Program Office
www.manufacturing.gov